Samela Stefane C. Silva
Gerlany F. S. Pereira

Education for Sustainable Development

Samela Stefane C. Silva
Gerlany F. S. Pereira

Education for Sustainable Development

Proposal to reduce food waste in the context of the Amazon region

ScienciaScripts

Imprint

Cover image: www.ingimage.com

This book is a translation from the original published under ISBN 978-3-330-20034-0.

Publisher:
Sciencia Scripts
is a trademark of
Dodo Books Indian Ocean Ltd. and OmniScriptum S.R.L publishing group

120 High Road, East Finchley, London, N2 9ED, United Kingdom
Str. Armeneasca 28/1, office 1, Chisinau MD-2012, Republic of Moldova, Europe
Printed at: see last page
ISBN: 978-620-8-27056-8

"If you have goals for a year. Plant rice. If you have goals for 10 years. Plant a tree. If you have goals for 100 years, then educate a child. If you have goals for 1000 years, then preserve the environment."

(Confucius)

DEDICATION

Samela Stefane Carvalho da Silva

To my parents Jedonias Bezerra da Silva and Maria Marcilene Carvalho da Silva.

Gerlany de Fatima dos Santos Pereira

To those who think and believe in differentiated education, capable of promoting the awareness we need to live in society. Much more than science educators, these professionals are educators for life. In this way, this book is dedicated to everyone who thinks about conserving humanity's greatest asset, which will only be possible if we respect and care for the world we live in!

ACKNOWLEDGMENTS

Samela Stefane Carvalho da Silva

I want to thank my GOD first of all, because he dreamed my dreams and made them come true, he blessed me and brought me this far. I know that I am nothing without his mercy and love;

With great love, I would like to thank my parents, who are the foundation of all the success in my life, who push me to succeed every day;

To my brother, Lucas Daniel, who stood by my side throughout the years of my degree;

My eternal gratitude to the wonderful people who crossed my path and who are no longer in my life today;

Finally, to my parents and friends who fought and took care of me, I share my joy with all of them.

Gerlany de Fatima dos Santos Pereira

Thank you to everyone who made this research possible!

SUMMARY

The aim of this study was to investigate the use of Education for Sustainable Development as an alternative to reducing food waste in a Full-Time State School in the State of Amapa, Brazil. This was a qualitative-quantitative study, which monitored the preparation and distribution of meals by weighing them. This took place in two stages, before and after the workshop. The workshop was on the theme of "Clean Plate", after which the questionnaire was applied. In the first stage, 27.64 kg of food was wasted after the workshop, while in the second stage, only 11.255 kg of food was wasted. It is understood that food loss was lower after the workshop, as it led school staff to reflect on the possibilities of making adjustments to food production.

Keywords: Education. Sustainable development. Waste. Food.

INDICE

1 INTRODUCTION

We are working here with the perspective of Education for Sustainable Development (ESD) as a possible solution to the problem of food waste. It is an education designed to prepare for the future. It aims to make human beings capable of facing today's main challenges, such as protecting the environment, respecting biodiversity and defending human rights (BURNETT, 2009).

In this way, it is believed that education from the perspective of sustainable development (SD) could be a way of combating ED. ESD enables every human being to acquire the knowledge, skills, attitudes and values needed to shape a sustainable future.

Brazil's diversity of natural resources makes it a key country in terms of environmental preservation and SD (UNESCO, 2015). Particularly in the Amazon region, due to its vast and rich biodiversity, support for ESD is becoming increasingly necessary. This is because the preservation/conservation of all forms of life and respect for them and the environment is something that should be disseminated from primary school to university classes.

Observing the context presented, the following research question arises: Can the use of ESD reduce DA in a Full-Time State School in the state of Amapa?

Thus, as a hypothesis, it is believed that if the school manages to work on ESD in order to sensitize students, teachers and the school community towards reducing AD, this will be an alternative for minimizing this problem, developing in students a differentiated view of the issue in question.

The aim of this research was to investigate the use of ESD as an alternative for reducing DA in a full-time state school in the state of Amapa. As well as, as specific objectives, to monitor the stages of preparation and distribution of meals in the School-Field of research to verify the DA, the development of a Thematic Workshop (OT) of ESD and DA with the students of the school in question.

Subsequently, to ascertain the implications of OT related to DA in the school surveyed, and to analyze whether ESD will be able to promote students' understanding of the importance of food and its conscious consumption.

2 FOOD AND SUSTAINABILITY: DIALOGICAL POSSIBILITIES

As generations change, their interests also change, intensifying the appropriation of natural resources to satisfy their desires at any given time. Therefore, future generations will be the heirs of everything that is done today (SIENA, 2008). Sustainability should be thought of as a way of not wasting existing resources so that the future of citizens is not compromised (DIAS, 2014).

An essential resource that has been produced on a large scale is food, on which human beings depend for daily quantities for their survival.

Gazzoni (2013) reports that it**'s not just a question of food, but of other resources that are invested in this productivity**, such as the large amounts of land that are cleared every day in order to grow more crops, which consequently results in a huge loss for ecosystems around the world. Another resource portrayed is water. Considered to be of paramount importance for human survival, it is used and wasted in alarming quantities, with 70% of all annual freshwater catchment being used for agriculture (GAZZONI, 2013).

It is also important to mention the use of energy. This has led to discussions about its sustainable use, as proper management of the natural resources that generate energy will ensure its durability for everyone and its extension to other generations. It is therefore essential to economize and not place an excessive burden on these resources (PIOLI, 2014).

We also need to think about population growth, which is expected to increase by more than 4 billion people by 2050. In percentage terms, to meet all this demand we would need to increase food production by up to 60%. However, if we reduced waste significantly to 50% there would be no need to exploit more land, what is produced and 30% more would feed everyone (GAZZONI, 2013), this would imply a reduction in the use of resources.

The surge in population growth and production in the different regions of the world will lead many countries to rely more and more on imports (UNEP, 2009). This immediately leads to greater use of energy, transportation and marketing of products. In order to reduce the environmental impacts caused by these processes, a movement is growing in Europe and the United States that focuses on valuing local food, i.e. consuming products obtained with local inputs, reducing the process between production and the table, and encouraging local producers (NITZKE et al., 2012).

The main function of food production is to meet the population's demand in terms of quantity and quality. On the other hand, the effects of high productivity have been discussed, such as the constant pollution of groundwater, soil and air. As a result of agricultural practices, there has been an increase in the emission of polluting gases such as methane, nitrous oxide and carbon dioxide, which increases the concentration of greenhouse gases in the atmosphere (MILIBAND, 2009).

Araujo and Pedrosa (2014) report that education is not the solution to the problems faced globally, but education itself transforms, liberates and through it a lot can be achieved. In this way, education becomes an ally in the path that has led to major changes in the world, generating critical citizens who are committed to making contributions that encourage a sustainable world.

As a result of living in a globalized age, information reaches everyone very quickly, and students are no strangers to this reality. As a result, students arrive in the classroom with the knowledge that society and the media offer them. Thus, they have prior knowledge[1] , which comes about through interaction with the environment in which they live and the conditions of that environment. It is at this point that the role of the teacher has a great influence on ESD, since they are the mediators of knowledge.

Hoffman (2000) reports on the importance of mediating action, understood as a constructivist stance[2] in education, in which the dialogic relationship, the exchange of discussions, the provocation of students, enables progressive understanding between teacher/student, thus advocating improvements in

---

[1] David Ausubel's Significant Learning Theory (SLT) states that the learning process can be facilitated by organizing teaching based on **students' prior knowledge** (PEREIRA; PADILHA, 2014a; PEREIRA; PADILHA, 2014b). Thus, the student's prior knowledge (concepts, propositions, principles, facts, ideas, images, symbols) is fundamental to this theory, since it is a determinant of the learning process (AUSUBEL; NOVAK; HANESIAN, 1980; AUSUBEL, 2003).

[2] It admits that knowledge is built up by the person themselves and is therefore neither transmitted nor revealed (CASTRO; CARVALHO, 1994).

education.

In order to achieve these improvements, teachers must be valued and qualified. Thus, when it comes to teacher training and its influence on the educational context, Krasilchik (1987, p. 47) talks about educational quality:

> Degree courses have been criticized for their ability to prepare teachers, making them capable of teaching good courses, according to the conceptions of what they aspire to in terms of training for science teaching; they have deficiencies in methodological areas that have extended to knowledge of the subjects themselves, leading to insecurity in relation to the class, the low quality of lessons and a close dependence on textbooks.

In this sense, we need to look at teacher training as something that is constantly under construction, with an initial and ongoing process[3] (MEDEIROS, 2010). Teachers need to be up-to-date on educational, social and technological issues. There is a need to fill gaps in initial training with continuing training, so that quality education can be achieved.

In view of this, ESD establishes direct links with teacher training and the need to promote it can be a strong ally in the training of critical educators. It is believed that it is through

[3]According to Libaneo (2004, p. 76) continuing education is a different way of looking at the professional training of teachers. It aims at personal and professional development through the involvement of teachers in the organization of the school, in the organization and articulation of the curriculum, in pedagogical-didactic assistance activities together with pedagogical coordination, in pedagogical meetings, in class councils, etc. The teacher is no longer just fulfilling the routine and carrying out tasks, without time to reflect on and evaluate what he or she is doing.

differentiated training that the socio-political and economic aspects related to environmental degradation will be worked on in order to form active critical citizens, stimulating effectively sustainable attitudes (VASCONCELOS; CONCEIQAO; FREITAS, 2012).

This research therefore advocates the existence of a dialogue between sustainable practices, ESD and the reuse of DA, while always respecting the environment and the resources it makes available to humanity. It is believed that only when man feels he belongs to the universe (CAPRA, 2006) will humanity be able to live in harmony with the planet. And all this will only be possible when education is taken seriously, respected and valued.

2.1 EDUCATING FOR SUSTAINABLE LIVING

Capra (2006, p. 13) in "Ecological Literacy", alludes to the web of life, the need to understand it and how interconnected we are with the environment. He says that "[...] the first step in this endeavor will have to be a fairly detailed knowledge of how nature sustains the web of life", from which point "we become aware of who we are, our role and how important we are in the process of the web of life".

For Capra (2006, p. 23) "The more we study the main problems of our time, the more we are led to realize that they cannot be understood in isolation. They are systemic problems, which means that they are interconnected and interdependent". Therefore, it is understood that sustainability is not just for one part, it goes further, involving the community as a whole. You stop being a system inserted in the environment and become part of it.

Also according to Capra (2006, p. 25), a reflection that should be recognized concerns the interdependence between all phenomena, the web of life shows that both individually and as a society "[...] we are all embedded in the cyclical processes of nature". In this way, one can grasp the ecological vision of human beings, which makes them think about how everything "fits into its natural and social environment".

Fritjof Capra (2006, p. 15) reinforces that education for sustainable living offers more possibilities than exist, because it creates and strengthens emotional bonds with nature, which makes "[...] our children become responsible citizens who are really concerned about the sustainability of life".

According to Gadotti (2008, p. 68), ESD is understood in this text to be "[...] concerned with learning attitudes, perspectives and values that guide and drive people to live their lives more sustainably" and that ESD "[...] is not only concerned with a healthy relationship with the environment, but with the deeper meaning of what we do with our existence, starting from everyday life" (GADOTTI, 2008, p. 66).

3 MATERIAL AND METHODS

This is a qualitative-quantitative study. According to Minayo (2010, p. 54) "the method has a fundamental function: to make it plausible to approach reality based on questions asked by the researcher". Quantitative and qualitative methods should therefore be used:

> [...] when developing a research proposal and carrying out the stages of a study, the researcher works with the recognition, convenience and usefulness of the available methods, in view of the type of information needed to fulfill the objectives of the work (MINAYO, 2010, p. 54).

Thus, in this study, questions relating to qualitative data were addressed. However, quantitative characteristics were also used in the stages of data organization, tabulation and analysis, using the phases of the statistical method of Zaros and Medeiros (2011), namely: Problem Definition, Research Planning, Data Collection, Data Critique, Data Presentation, Analysis and Interpretation.

3.1 CHARACTERIZATION OF THE STUDY AREA

The data for this study was collected at the State Full-Time School, located at Rodovia Duca Serra, 1753 - Km 06, in the Coragao District, Macapa - Amapa.

With a total of 379 students, distributed over 2 shifts, namely: morning[4] (with 161 students) and afternoon (218). In this study, only the students of Elementary School II were included, as this is the target audience of the Natural Sciences teachers.

3.2 THE STUDY PERIOD

This research took place in 2016, starting in April and ending in January 2017.

3.3 THE RESEARCH SUBJECTS

Elementary school students at a full-time state school in the state of Amapa.

TABLE 1. Distribution of students by class.

Classes	Number of students
6TH GRADE - A	25
6TH GRADE - B	26
6TH GRADE - C	27
YEAR 7 - A	25
YEAR 7 - B	27
721 (7ª SERIES)	27
722 (7th SERIES)	24
821 (8th SERIES)	37
Total number of students	**218**

Source: Field research.

[4] With 2 1st grade classes, 1 2nd grade class, 1 3rd grade class, 2 4th grade classes and 3 5th grade classes.

3.4 DATA COLLECTION

In order to investigate DA at the school under investigation, the stages of meal preparation and distribution were monitored by means of weighing - the First Food Weighing Stage. To this end, four days of weighing were carried out at this stage of the research.

The meals were weighed before being served to obtain the **preparation yield** (PR). Subsequently, the amount of leftovers was weighed again after the meals were finished (**clean leftovers** - SL and **leftovers** - RT).

The value of the leftovers, taken from the preparation yield, will be the equivalent of the amount of **food consumed -** AC. According to Abreu; Spinelli and Zanardi (2003), clean leftovers are foods that are ready but have not been distributed.

During data collection, the "field diary" was used to make notes on the observations of situations considered important, experienced during the study and which were not included in the data collection instruments recommended in the presentation of the research project. Lewgoy and Reidel (2009) point out that the contents described in the diary begin by describing the concrete, then proceed to abstraction and generalization, returning to the concrete, now understood as the concrete thought.

After the first stage of weighing the food, the students were given a workshop on ESD and AD. After OT, the **Second Food Weighing Stage** took place, in which the implications of OT on DA were investigated. The data was obtained through new weighings, using the same methodology as the first stage, over 4 weighing days.

Through this data, it was possible to ascertain whether or not the ESD assumptions had positive implications for the understanding of the importance of food and its conscious consumption.

Finally, the Perception Questionnaire was used to ascertain the students' knowledge of ESD and AD.

3.5 THE THEMATIC WORKSHOP

In order to develop the OT on ESD and LD with the students at the school under investigation, a plan was drawn up with the supervisor for the activity to be carried out at the school. This was applied after the first stage of data collection.

The theme of the Thematic Workshop on ESD and DA was "Clean Plate". The resources used were lectures and dialogues with multimedia slide shows, using photography as the main resource. This resource is an instrument that facilitates the understanding of social, economic and environmental aspects, among others (SEVERINO, 2010; JUSTO, 2003). Thus, photography can be understood as an infinite source of data, facts and information that interact in the process of materialization, which facilitated the teaching and learning process.

In this way, we looked at the resources invested in food production (water, energy, soil, etc.), the stages in which waste occurs, the discrepancies between the amount of food produced and the large population group that suffers from food shortages, and the consequences of AD.

3.6 DATA ORGANIZATION, ANALYSIS AND INTERPRETATION

The data from the First and Second Food Weighing Stages were organized into tables and analyzed comparatively. The perception questionnaire was analyzed using content analysis, according to Bardin (2009), with "categorization" and "inference" being the most important analysis techniques.

3.7 ETHICAL ASPECTS

Data collection took place after the students' parents had signed an informed consent form. The signing of the consent form was preceded by a presentation and discussion with the participants' guardians about the research objectives, methodology, benefits and probable risks, such as embarrassment at taking part in the activities proposed in the course, as well as embarrassment at being filmed, photographed or having their speech recorded. During this process, the parents of the students asked for clarification about the aspects of the research and the inviolability of their identities.

4 RESULTS AND DISCUSSIONS

4.1 MONITORING THE PREPARATION AND DISTRIBUTION OF MEALS TO CHECK FOR FOOD WASTE

4.1.1 First stage of weighing the food

TABLE 2: Values from four days of weighing in the first stage of food weighing.

Snacks Served	Weighing days	Preparation yield	Leftovers Clean	Leftovers	Meals served	Meal per Student
Macaroons Sausage	1st day of weighing	24.565 kg	5 kg	1,54 kg	98 meals	200 g
Vatapa	2nd day of weighing	37, 550 kg	15, 150 kg	0 kg	116 meals	193g
Rice		13 Kg	0 Kg	0 kg		110 g
Porridge Tapioca	3rd day of weighing	15,600 kg	0 kg	0 kg	70 meals	220 g
Nescau	4th day of weighing	26, 600 kg	5.8 kg	0 kg	84refeipoes	250 g
Cookie		2 kg	150 g	0 kg		22 g
	Total:	**119, 315 kg**	**26,100 kg**	**1,54 Kg**	**368 meals**	**995 g**

Source: Field research.

From the first day, it can be seen that of the 24.565 kg of food produced for the students' consumption, when the RT and SL values are added together, 6.54 kg of food was wasted. Given this figure, and taking into account the amount of food per student (200g), it would be possible to serve 32 more meals. In the researcher's opinion, this indicates a huge unnecessary waste of food.

To make up the menu for the second day of weighing, rice and vatapa had to be produced. The 15,150 kg of SL would be enough to serve another 49 meals. It is interesting to note that there was no RT. This may be due to the students' preference for the menu of the day. The saying "tasty food doesn't go to waste" was present in the students' speech.

On the third day of weighing, there was no RT or SL. This shows that the quantity produced was proportional to the number of meals, and that the students approve of and like the food served. This meant that consumption was total.

For the fourth weigh-in day, the snack served was iced Nescau and cookies. There was no RT on this day. However, a significant amount of Nescau SL was obtained, which would be enough to serve approximately 23 more meals. In terms of the amount of cookies, 6 more people could have been served with the amount of SL. This shows the disproportionate amount of food prepared.

4.1.2 Second stage of food weighing

It is important to mention that this stage of the research was carried out after the workshop. For another 4 days of weighing, the results are shown in Table 3 below.

TABLE 3: Values from four days of weighing in the second stage of food weighing.

Snacks Served	Weighing days	Preparation yield	Leftovers Clean	Leftovers	Meals served	Meal per Student
Tapioca porridge	5th day of weighing	15, 600 kg	1,100 kg	0 kg	65refeigoes	222 g
Maria Izabel	6th day of weighing	24,400 kg	1,600 Kg	1,800 kg	105 meals	200 g
Vatapa **Rice**	7th day of weighing	22,400 kg 16kg	0 kg 3 kg	980 g 0 g	115 meals	194 g 110 g
Nescau **Cookie**	8th day of weighing	26.6000 kg 2 kg	1,235 kg 0 Kg	0 kg 0 kg	101refections	250 g 22 g
	Total:	**107 kg**	**6, 935 Kg**	**4,320 Kg**	**386 meals**	**1.001 kg**

Source: Field research.

Compared to the third day of weighing of the first stage of food weighing, it is noticeable that fewer students were interested when the snack offered was porridge. Similarly, on the fifth day of weighing, there were no leftovers. However, because fewer meals were served, 1,100 kg of SL was generated, which would be enough to serve 5 more meals, taking into account 222g per student.

For the sixth day of weighing, it can be seen that of the 24,400 kg of food produced for the students' consumption, there was a SL quantity of 1,600 kg of food. From RT, a figure of around 1,800 kg was obtained. The sum of the amounts wasted is 3,400 kg, where 17 more meals could have been served.

On the seventh day of weighing, rice and vatapa had to be produced to make up the menu to be served. Once again, it should be borne in mind that because these foods are prepared separately, they were not produced in proportional quantities.

However, if we compare this with the second day of weighing, which also included rice and vatapa, we can see that the amount of vatapa produced was lower. This may have been due to the fact that the school staff were shown the values from the first stage of weighing the food, as they went from a value of 15.150 kg of FS (second day of weighing) to just 3 kg of FS on the seventh day of weighing.

For the eighth day of collection, a comparison is made with the fourth day of the first stage of food weighing, where Nescau was served with cookies, and there was no RT of any of the items served. Here there was a greater number of students who showed interest in the meal after the workshop, and the greater number of meals served reduced the number of FS from 5.8 kg on the fourth day to just 1.235 kg of FS, which could still have been served 5 more meals.

4.2 DEVELOPMENT OF A THEMATIC WORKSHOP ON EDS AND FOOD WASTE WITH STUDENTS AT A FULL-TIME STATE SCHOOL IN THE STATE OF AMAPA

There are many approaches that contribute to the process of teaching and learning natural sciences. Among them, OT stands out as a facilitating tool for integrating different areas of knowledge. With this in mind, we thought about developing a TO to approach ESD as an alternative to DA.

The main interest with OT was to show students the main resources involved in food production, how they are treated, their use, the effects they have on the environment, the long journey food takes to reach the consumer's table, how much food is produced and the differences between food production figures and world hunger.

The audience for the OT was 110 people, 98 of whom were students. The workshop took place in the school cafeteria, as it could accommodate all the students sitting down in the afternoon. Due to the number of students, we used the tables that were available to divide them into groups, so that it would be easier to listen to one group at a time and their opinions throughout the activity.

Images were used to give the students a better understanding of the issues at hand, because they allow for greater communication. This allowed them to have their own interpretations and feel part of the context being worked on.

The OT was conducted in this way, as it is believed that photography is a tool that has been used in the teaching and learning process, and is considered "[...] an instrument of great pedagogical importance and often essential for various areas of teaching" (BORGES; ARANHA; SABINO, 2010, p. 150).

Spencer (1980) considers that it contributes decisively to theoretical research, artistic and cultural manifestations and acts as an effective coadjuvant in scientific and technological discoveries. What's more, it contributes to science by providing us with a kind of synthetic eye - "an impartial and infallible retina" - capable of converting into visible records phenomena whose existence we would not otherwise have known or suspected (SPENCER, 1980). This was combined with the use of photographs/images as the guiding methodological resource for OT.

Before discussing the specific content of the workshop, the students were asked the following questions: What is sustainability? Do all people need food? Why do people go hungry? What resources are used to produce food? Why shouldn't we spoil food?

Each group then received a different picture. They contained records of the waste of natural resources in the production of food, the selection of food, the journey to the distribution points (supermarkets, restaurants, warehouses, etc.), the waste in households and the amount of food that ends up in the garbage.

The students then expressed their opinions about what they saw in the images, raising questions and answering their doubts. Some of the images portrayed a little of their daily experiences. In this way, they led the students to greater reflections, because according to Vygotsky (1994) they use and activate memory, where they update their previous knowledge in order to expand and transform their opinions.

At the end, the questions asked at the beginning of the OT were recalled, and it was now clear that the students no longer saw these questions as something unknown, but that they were able to answer them within their means.

At the end of the workshop, the images given to the groups were used to make a panel, where they organized the stages of AD and the inverted resources in productivity according to the content of the workshop.

4.3 INVESTIGATION OF THE WORKSHOP'S IMPLICATIONS FOR FOOD WASTE AT THE STATE FULL-TIME SCHOOL INVESTIGATED

It can be seen that OT had a positive impact on the school, as the students and the school's Pedagogical Coordination were mobilized to find out how much was being produced and to reduce AD. The first stages of weighing the food resulted in 27.64 kg of food being wasted between SL and RT. This figure corresponds to just four days of weighing.

It is important to note that of the totals shown above, SL has 26.1 kg and RT 1.54 kg. In order to avoid these high values, it is important to involve the entire staff of the institution, so that together they can draw up leftover control parameters, and also to train and sensitize the staff and, whenever possible, prepare the food in stages, according to demand (AUGUSTINI et al., 2008).

Right after the OT, the values from the first weighings were presented to the school staff, so that it was possible to readjust the amount of food produced in proportion to the number of students who use the school meals.

Now with food production proportional to the number of students, the DA was reduced, where in the second stage of weighing the food, only 6.935 kg of food was obtained in SL and 11.255 kg in RT, which generates a total of 11.255 kg of DA. This could still be improved with future support involving EDS. However, it is already considered that TO, in general, had a positive impact on the reduction of DA in the school investigated.

4.4 ANALYSIS: EDS AND ITS ABILITY TO PROMOTE UNDERSTANDING OF THE IMPORTANCE OF FOOD AND ITS CONSCIOUS CONSUMPTION AMONG STUDENTS AT THE INESTIGADA FULL-TIME STATE SCHOOL

The students at the Full-Time State School under investigation come from a background of intense poverty, many of whom live near the old Santana dump and in other adjacent neighborhoods that reflect poverty as their main problem. This can be seen, according to Bezerra (2009, p. 104):

> Through the initial data collection, I was able to understand that approaching the meanings of lunch could not happen without referring to the problem of hunger, not only because the school is located in a space marked by poverty, but because this issue is present in the discourse of the subjects involved in the research (BEZERRA, 2009, p. 104).

These problems are clearly perceptible in the students' speeches. According to Cardoso (2006), perception is a configuration and organization of elements that the human mind integrates into its past experiences, linking and unifying them, that is, filtering them through the factors of meaning that language and each person's cultural references have already created. With this in mind, a Perception Questionnaire on aspects related to AD was administered to the subjects surveyed here (90 students).

In the application, question 1ª asked: **What do you think is the importance of food for human beings?**

The categories for this question are as follows: 46.66% of the students believe that "food is of the utmost importance, as it is a question of survival", 37.77% believe that "food guarantees strength, provides energy and makes people healthy" and 15.55% believe that "food is a means of facilitating learning".

When the above categories are analyzed in detail, they are very similar to Pedraza's (2004, p. 7) thinking, when he alludes that "The normal functioning and structure of every living organism is determined, among other factors, by the adequate supply of nutrients and energy that allows for a satisfactory nutritional state, which can ultimately constitute one of the most elementary rights of man", which is: food. Because it is a "matter of survival", "guarantees strength" and "provides energy".

Taking into account the importance of what the students have said, in fact all human beings need food, but its importance is highlighted when it is an agent that facilitates the learning process. Moyses and Collares (1997) exemplify more clearly that hunger is the basic need for food, but when it is not met, it reduces the availability of any human being for daily activities as well as intellectual activities. This is perfectly in line with the last category mentioned above.

The school investigated has the possibility of offering school meals to all students, but not everyone is interested. The proportion of pupils who show an interest in school meals is made up of hungry pupils who don't have the resources to buy other food, as it is common for schools to offer a variety of snacks. These students who make use of school meals don't actually spoil them, because their importance of food is directly linked to the scarcity of food in their homes and the hunger they feel.

Food is a basic need, a human right and, at the same time, a cultural activity, permeated by beliefs, taboos, distinctions and ceremonies (PEDRAZA, 2004). From the author's perspective, eating is not just about incorporating important nutritional elements into our bodies, it is first and foremost a social act and, like any relationship between people, it involves coexistence, differences and expresses the world of need, freedom or domination. A group's eating patterns support its collective identity, position in the hierarchy and social organization, but certain foods are also central to individual identity (SUELI, 2001).

Regarding the importance of non-AD, the students' understanding can be seen in question 2ª : **Do you agree with the enormous waste of food in the world? Why?**

Based on this question, the students' answers were grouped into the following categories: 54.44% of the students "don't agree with DA, because they understand that there are a lot of people who go hungry", 33.33% "don't agree with DA, because when they do this they are also throwing away other resources from our planet" and 12.22% of the students "don't agree with DA, because it harms the environment by causing damage".

When students report in one of the categories that they "don't agree with the DA, because they understand that there are a lot of people who go hungry", it is clear that they understand that high food productivity does not provide sufficient conditions to prevent part of the population from going hungry. Hunger is often associated with a lack of equal distribution of income (PEREIRA, 2012; PEREIRA; RIBEIRO; FREITAS, 2014). However, hunger is also linked to other factors that need to be taken into account, such as the legal, social and economic structure of society (HOFFMANN, 1995).

In view of the statements in the categories that show that students do not agree with AD, because its waste intensifies the loss of many natural resources, as well as accentuating the damage to the environment, they show that all attitudes, whether conscious or unconscious, end up having repercussions, which can be positive or negative.

Thus, it is necessary to propose the inclusion of ecological care in daily practices, which is "[...] understood as a set of skills to contextualize and produce a broad and complementary awareness" (MORIN, 2002). These can be "individual discipline to defend the environment and collective awareness of the need to provide future generations with quality living conditions". To reach this level of awareness, education is the main ally, in its broadest sense, and specifically ESD, which tends to be a natural consequence of the broader process (ZULAUF, 2000).

The students were asked about their understanding of the factors that lead to **food waste,** starting with question 3ª : **What do you think leads to food waste?**

The students' responses were grouped into the following categories: 60% of the students justified the DA by " lack of awareness of the importance of the subject ".

food" and for 40% also caused by "high consumption unnecessary".

The two categories above are strongly interlinked; one is seen as a consequence of the other. For when they express the idea that DA 32 and generated by the "lack of awareness of the importance of food", justifies the fact that people don't realize or understand, according to Cavalcanti (2004), that "nature is our primordial and irreplaceable source of life and at the same time acts as the ultimate drain of dirt".

The author goes on to clarify that an economic level has been reached that only "promotes the wasteful use and depletion of natural resources, favors the unsustainable reproduction of consumption/destruction patterns, and provides no effective benefit in terms of the human well-being of the disadvantaged masses" (CAVALCANTI, 2004).

And as a consequence of the lack of sensitivity to natural resources, "unnecessary high consumption" is one of the causes of AD. Campbell (2001) justifies this in the following way, that today's population, characterized by extreme consumption, is always insatiable and mainly unsatisfied "where a need that is first satisfied almost automatically generates another need, in a cycle that never ends, in a *continuum* where the end of the consumerist act is the desire for consumption itself".

All the questions above led the students to reflect on the concepts covered in the OT. The categories of answers show how enlightening the perception questionnaire was. Finally, in question 4ª , the following question was asked: **What do you understand by sustainability?**

The students' answers fell into the following categories: 65.55%

answered that "sustainability is about attitudes that guarantee quality of life, good for everyone and the preservation of nature" and 34.44% of the students also answered that "sustainability is about human attitudes and actions, both economic and environmental, that ensure that natural resources are used correctly".

It can be seen that the students' view of the concepts concerning sustainability brings a relationship of balance between man and nature, as they clearly express this statement, where "sustainability are attitudes that guarantee quality of life, the good for all and the preservation of nature".

These attitudes can be affirmed by Diniz and Bermann (2012), who say: "The issue of equality between generations from the perspective of sustainability means that each generation should have the same well-being, or the same equal opportunities, as the others", thus seeking to improve attitudes that go beyond individualistic barriers.

According to Elkington (1994), achieving sustainability requires a balance between three important pillars: environmental, economic and social. Elkington's vision and support for the subsequent category "sustainability are human attitudes and actions, both economic and environmental, which provide for the correct use of natural resources", shows that for the students sustainable attitudes must be taken by everyone, that actions must be planned and put into practice.

Given all the questions raised above, we can see the fundamental role of education being put into practice and achieved. According to Nascimento and Hetkowski (2009), education is "[...] an intentional, conscious process, based on valuing life and seeking to guide people towards knowledge of themselves [...]", leading people to attitudes that have a positive influence on changing collective values and behaviors (JACOBI, 1997).

5 FINAL CONSIDERATIONS

The aim of this study was to investigate the use of ESD as an alternative for reducing AD in a full-time state school in the state of Amapa. The results showed that the scenario investigated is the scene of problems related to a precarious socio-economic condition on the part of the subjects investigated. This shows that the educational issue (the search for knowledge) is not the only "attraction" of the school.

In the area where the research was carried out, it was seen that pupils also look to school as a means of satisfying their food needs, since school meals end up making up for the shortfall that many of these pupils have and are an incentive for them to stay in school.

It turned out that, from this perspective, OT proved to be an enlightening way of understanding ESD and how much it can influence the reuse of DA, as it led the employees of the school under investigation to develop attitudes that were fundamental to achieving a lower level of food waste. Because there would be no point in students consuming food without wasting it if management produced too many school meals.

Thus, combined with the socio-economic condition of the students and the school community and knowledge of ESD assumptions, it can be said that yes, OT has had an effect on AD.

There is a need for more studies on ESD as an alternative to AD, or at least to look for viable solutions. Given that we live in a society that values consumerism and sometimes (for the most part) forgets that it belongs to the universe, and the care that needs to be taken if humanity is to continue to exist on this planet.

But for this to be possible, a real change in attitudes, values and perceptions is more than necessary, and ESD can contribute to this.

REFERENCES

ABREU, E.; SPINELLI, M.; ZANARDI, A. **Gestao de unidades de alimentaria e nutrigao**: um modo de fazer. Sao Paulo: Editora Metha, 2003.

ARAUJO, M. F. F.; PEDROSA, M. A. Teaching Science from the perspective of sustainability: barriers and difficulties revealed by biology teachers in training. **Educar em Revista**, n. 52, p. 305-318, 2014.

AUGUSTINI, V.C.M.; KISHIMOTO, P.; TESCARO T.C.; ALMEIDA, F.Q.A. Avaliapao do indice de resto-ingesta e sobras em unidade de alimentaria e nutripao de uma empresa metalurgica na cidade de Piracicaba/SP. **Rev Simbio-Logias**, v. 1, n. 1, p. 99-110, 2008.

AUSUBEL, D. P. **Acquisition and retention of knowledge**: a cognitive perspective. Lisbon: Platano, 2003. 219 p.

AUSUBEL, D. P.; NOVAK, J.D.; HANESIAN, H. **Psicologia Educacional**. Rio de Janeiro: Interamericana, 1980. 626 p.

BARDIN, L. **Content analysis**. Lisbon: Edipoes 70, 2009.

BEZERRA, J. A. B. **Food and school: meanings and curricular implications of school meals.** Rev. Bras. Educ. [online]. 2009, vol.14, n.40, pp.103-115.

BORGES, M. D.; ARANHA, J. M.; SABINO, J. Nature photography as a tool for environmental education. **Ciencia & Educagao**, v. 16, n. 1, p. 149-161, 2010.

BURNETT, N. **UNESCO: What is Education for Sustainable Development**? 2009. Available at : <https://criaeinova.wordpress.com/2009/06/15/unesco-o-que-e- educacao-para-o-desenvolvimento-sustentavel>Accessed on: September 10, 2015.

CAMPBELL, C. **Romantic ethics and the spirit of modern**

consumerism. Rio de Janeiro: Rocco, 2001.

CAPRA, F. **The web of life**: a new scientific understanding of living systems. Sao Paulo: Cultrix Publishing House, 2006.

CARDOSO, J. F. **Sensagao e Percepgao**. jul., 2006. Available at: <http://moranapsicologia.blogspot.com.br/2006/07/sensao-e-percepo-i.html>. Accessed on: January 30, 2016.

CASTRO, R.; CARVALHO, A. M. P. The historic approach in teaching: Analisis of an expirience. **Educational Research**, v. 23, n. 7, p. 4, 1994.

CAVALCANTI, C. An attempt to characterize ecological economics. **Ambient. Soc.**, Campinas, v. 7, n. 1, jun., 2004.

DIAS, N. A. **Sustainability in food and nutrition units: challenges for nutritionists in the 21st century**. Monograph. Degree in Nutrition. Federal University of Juiz de Fora - UFJF, Juiz de Fora - MG, 2014. Available at : <http://www.ufjf.br/gradnutricao/ensino/tcc/trabalhos/2014-2/2014-2-2/>. Accessed on: August 9, 2015.

DINIZ, E. M.; BERMANN, C. Green economy and sustainability. **Estudos Avangados**, v. 26, n. 74, p. 323- 330, 2012.

ELKINGTON, J. Towards the sustainable corporation: Win-win-win business strategies for sustainable development. **California Management Review**, v. 36, n. 2, p. 90-100, 1994.

FOLEY, J. Five steps to feeding the world. **National Geographic magazine**, May 2014 . Available: <http://viajeaqui.abril.com.br/materias/national-geographic-maio-de- 2014-edicao-170>. Accessed on: August 8, 2015.

GADOTTI, M. **Educar para a sustentabilidade**: uma contribuigao a decada da educação para o desenvolvimento sustentavel. Sao Paulo: Editora e livraria Instituto Paulo Freire, 2008.

GAZZONI, D. L. More food, with sustainability. **Cultivar magazine**, p. 37,

May 2013. Agronegocios column. Available at: <www.revistacultivar.com.br>. Accessed on: August 10, 2015.

GUILHERME, N. C.; CASTRO, A. M.; TORRES, V. Q.; TUCHERMAN, M. **Reducing food waste**. Hospital Israelita Albert Einstein - SP - Brazil. Case Study. April 14, 2014.

HOFFMAN, J. **Avaliação Mediadora**: Uma Pratica da Construgao da Pre-escola a Universidade. 17 ed. Porto Alegre: Mediagao, 2000.

HOFFMANN, R. Poverty, food insecurity and malnutrition in Brazil. **Estudos Avangados**, Sao Paulo, v. 9, n. 24, 1995.

JACOBI, P. Urban environment and sustainability: some elements for reflection. In: CAVALCANTI, C. (Org.). **Meio ambiente, Meio ambiente, desenvolvimento sustentavel e politicas publicas**: sustainable development and public policies. Sao sustainable development and public policies. Paulo: Cortez Editora, 1997.

JUSTO, C. S. **The boy photographers and the educators**: living on the street and in the Casa Project. Sao Paulo: UNESP, 2003. 237 p.

KRASILCHIK, M. **O professor e o curriculo das ciencias**. Sao Paulo: EPU: Editora da Universidade de Sao Paulo, 1987. 80p.

LEWGOY, A. M. B.; REIDEL, T. **Field Diary: What is it? What is it for? How to prepare it?** 2009. Available at: http://www.ufrgs.br/psicologia/graduacao/servico-social/comgrad/comissãoo-de-estagios/DiariodecampoModelo.pdf

LIBANEO, J. C. **Organização e g**estao **escolar**: teoria e pratica.5ed. Goiania: Editora Alternativa, 2004.

MEDEIROS, L. G. **Metodologia e Instrumentagao para o ensino de Ciencias Naturais**. Cadernos Cb Virtual 5. Joao Pessoa: Ed. Universitaria, 2010.

MILIBAND, E. C.: The turning point for the climate. **Citizenship and**

environment magazine. Camara Cultura, n. 23, p. 32, 2009.

MINAYO, M.C.S. **O desafio do conhecimento**: Pesquisa Qualitativa em Saude. 12 ed. Sao Paulo: Editora Hucitec, 2010.

MORIN, E. **Os sete saberes necess necessárias a educagao do futuro**. 5 ed. Sao Paulo: Cortez; 2002.

MOYSES, M. A.; COLLARES, C. Malnutrition, school failure and school meals. In. PATTO, M. H. (Org.) **Introdugao a psicologia escolar**. 2 ed. Sao Paulo: Casa do Psicologo, 1997.

NASCIMENTO, A. D. (Org.); HETKOWSKI, T. M. (Org.) **Educagao e contemporaneidade**: pesquisas cientificas **e** tecnologicas. Salvador: EDUFBA, 2009.

NITZKE, J. A.; THIS, R.; MARTINELLI, S.; OLIVERAS, L. Y.; RUIZ, W. A.; PENNA, N. G.; NOLL, I. B. Food safety - a return to the origins? **Brazilian Journal of Food Technology**, IV SSA, May, p. 2-10, 2012. Available at : <http://www.scielo.br/pdf/bjft/v15nspe/aop_bjft_15e0102.pdf>. Access on: Aug. 10, 2015.

PEDRAZA, D. F. Padroes Alimentares: da teoria a pratica - o caso do Brasil **Mneme - Revista Virtual de Humanidades**, n. 9, v. 3, jan./mar.2004. ISSN 1518-3394

PEREIRA, G. F. S. **Appropriation of Scientific Knowledge: an approach to Transgenic Foods**. 120 f. Dissertation (Master's Degree in Science and Mathematics Education). Postgraduate Program in Science and Mathematics Education (PPGECM), Institute of Mathematical and Scientific Education (IEMCI), Federal University of Para (UFPA), 2012.

; RIBEIRO, E. O. R.; FREITAS, N. M. S. **Appropriation of Scientific Knowledge**: an approach to Foods Transgenics. 1. ed. Saarbrucken, Germany: Verlag-Novas Edigoes Academicas, 2014. v. 1.96p.

PADILHA, E. C. P. The importance of prior knowledge in science teaching in the light of David Ausubel's Significant Learning Theory. In: Proceedings... V ENCONTRO NACIONAL DE APRENDIZAGEM SIGNIFICATIVA-V ENAS, Belem-PA, August 2014a.p. 250-259.

The role of prior knowledge in the process of learning meaningfully: approaches to science teaching. In: **Anais...** V NATIONAL MEETING ON LEARNING
SIGNIFICATIVA-V ENAS, Belem-PA, August 2014b. p. 966-974.

PIOLI, C. Considerations for preparing a sustainable menu. In: SILVA, S. M. D. S. **Cardapio**: guia pratica para elaboragao. Sao Paulo: Roca, 2014. p. 365-373.

SEVERINO, F. E. S. The pedagogical mediation of photography in the teaching of transversal themes. **Educapao & Linguagem**, Sao Paulo, v. 13, n. 21, p. 175-188, jan./jun., 2010.

SIENA, O. Metodo para avaliar desenvolvimento sustentavel: tecnicas para escolha e ponderapao de aspectos e dimensões. **Produpao**, v. 18, n. 2, p. 359-374, 2008.

SPENCER, D. **Color Photography in Practice**. 2. ed. London: Iliffe & Sons, 1980. 244 p.

SUELI R, T. **Malnutrition and Obesity:** Contradictory Faces in Poverty and Abundance. Maternal and Child Institute of Pernambuco. Serie: Publicapoes Cientificas do Instituo Materno Infantil de Pernambuco (IMIP), n. 2. Recife. 2001

UNESCO. **Education for Sustainable Development in Brazil**. 2015. Available at: <http://www.unesco.org/new/pt/brasilia/natural-sciences/education-for-sustainable-development/>. Accessed on: September 10, 2015.

UNITED NATIONS ENVIRONMENT PROGRAM - UNEP. The environmental food crisis. **Citizenship and Environment Magazine**. Camara Cultura, n. 23, p. 06-11, 2009.

VASCONCELOS, E. R.; CONCEIQAO, L. C. S.; FREITAS, N. M. S. Ideas about sustainable development: science education and the CTS approach, possible articulations. **Rev. eletronica Mestr. Educ. Ambient**. v. 28, January to June 2012.

VYGOTSKY, L. S. **The social formation of the mind.** 5 ed. Sao Paulo: Martins Fontes, 1994.

ZAROS, L. G.; MEDEIROS, H. R. **Bioestatistica**. 2 ed. Natal-RN: EdUFRN, 2011.

ZULAUF, W. **The environment and the future**. Estudos Avanpados. v. 14, n. 39, Sao Paulo, 2000.

APPENDIX A

PERCEPTION QUESTIONNAIRE

Researcher: Samela Stefane Carvalho da Silva
Research: "Education for Sustainable Development: a alternative for reducing food waste"
Supervisor: Gerlany de Fatima dos Santos Pereira

1. How important do you think food is for human beings?

2. Do you agree with the enormous waste of food in the world? Why?

3. What do you think leads to food waste?

4. What do you mean by sustainability?

Printed by Books on Demand GmbH, Norderstedt / Germany